Starting an E-Waste Recycling Business Guide for Beginners

Roggie Clark

Continue Learning At Our Website: Roggieclarkbooks.com

Continue exploring educational resources, tools, books, workbooks, and learning materials designed to help readers continue building knowledge and practical skills. Visit: www.roggieclarkbooks.com

Inside you'll find:

- Educational worksheets
- Beginner-friendly guides
- Printable activities
- Companion materials for books and courses

A growing library of educational content for readers, collectors, learners, parents, teachers, and lifelong students.

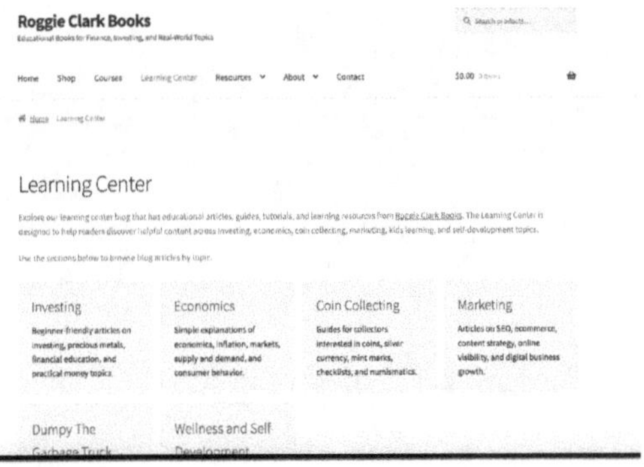

The Strategic Guide to U.S. Mint Gold and Silver Coin Investing

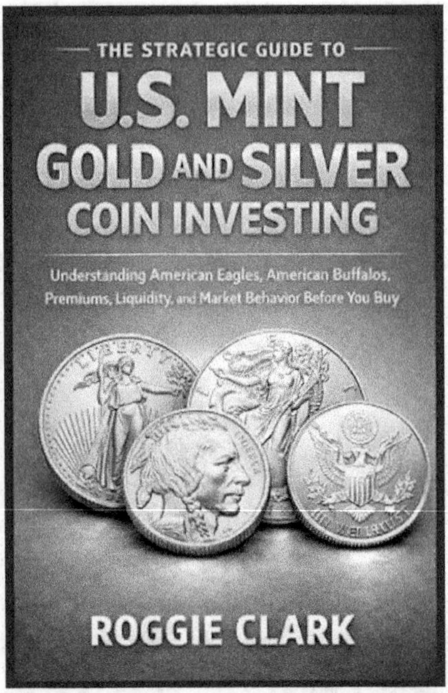

Rather than offering a general introduction to precious metals, this guide examines how American Gold Eagles, American Gold Buffalos, and Silver American Eagles behave in real investor markets.

https://www.amazon.com/Strategic-Guide-U-S-Silver-Investing/dp/B0GQKJF9K6

Copyright

Title: Starting an E-Waste Recycling Business Guide for Beginners
Author: Roggie Clark

Cover Design: Roggie Clark
Editing: Renee' Clark

Printed in the United States of America

Legal Disclaimer

The information presented in this book is provided solely for general educational and informational purposes. It is intended to help readers better understand electronic waste, electronics refurbishment, component recovery, material recycling, and the operation of an e-waste recycling business. It should not be interpreted as legal, financial, tax, environmental, engineering, electrical, electronics repair, recycling, data security, hazardous waste, or professional business advice.

The author and publisher make every effort to provide accurate and current information. However, laws, regulations, environmental requirements, hazardous waste policies, transportation rules, electronic recycling standards, data privacy laws, workplace safety requirements, and business licensing requirements vary by country, state, province, municipality, and organization, and they may change over time. Readers are solely responsible for researching and complying with all applicable laws, regulations, permits, certifications, and licensing requirements within their jurisdiction.

This book promotes only **legal and ethical methods** of collecting, transporting, refurbishing, dismantling, recycling, and selling electronic equipment. Nothing contained in this publication should be interpreted as encouraging theft, trespassing, unauthorized collection of electronic equipment, improper disposal of hazardous materials, violations of privacy laws, intellectual property infringement, unsafe recycling practices, or any other illegal or unethical activity. Always obtain permission before collecting, transporting, repairing, dismantling, or selling electronic equipment that belongs to another individual, business, or organization.

This book discusses electronics testing, refurbishment, component harvesting, data destruction, material recovery, and recycling processes solely to provide readers with a general understanding of how the e-waste recycling industry operates. These discussions are educational in nature and should not be interpreted as step-by-step repair instructions, technical repair manuals, electrical engineering guidance, or guarantees that any particular electronic device can be safely repaired, refurbished, or returned to service.

Electronic equipment may contain hazardous materials, including rechargeable batteries, mercury-containing lamps, capacitors, lead-containing components, toner cartridges, and other substances that require proper handling and disposal. Some electronics may also contain energized electrical components capable of causing electric shock, fire, burns, or other injuries if handled improperly. Appropriate personal protective equipment, safe work practices, proper ventilation, and manufacturer safety recommendations should always be followed when handling electronic equipment.

Data security is an important responsibility within the e-waste recycling industry. Computers, laptops, servers, smartphones, storage devices, and other electronic equipment may contain confidential personal, financial, medical, or business information. Readers are responsible for complying with all applicable privacy laws, data protection regulations, contractual obligations, and accepted industry standards regarding data sanitization and secure destruction of storage media.

References to businesses, manufacturers, software, tools, organizations, recycling facilities, government agencies, products, services, or industry standards are provided solely for educational purposes and do not constitute endorsements or recommendations. Business practices, acceptance requirements, pricing, regulations, certifications, and operating procedures vary among organizations and may change without notice.

Any discussion of business opportunities, profitability, resale values, component values, precious metal recovery, recycling revenue, operating costs, or financial performance is provided for illustrative purposes only and should not be interpreted as a guarantee of future income or business success. Actual results vary depending on market conditions, local competition, customer demand, operating expenses, business experience, regulatory requirements, equipment costs, and numerous other factors beyond the control of the author and publisher.

Readers are encouraged to seek qualified professional advice whenever legal, financial, tax, engineering, environmental, or technical guidance is needed. Always follow applicable safety regulations, manufacturer recommendations, and local laws when handling, transporting, refurbishing, dismantling, storing, or recycling electronic equipment.

By using the information contained in this book, you acknowledge that you assume full responsibility for your own decisions and actions. The author and publisher disclaim all liability for any loss, injury, illness, death, property damage, data loss, environmental damage, financial loss, legal consequences, regulatory penalties, business interruption, or other claims arising directly or indirectly from the use or misuse of the information presented in this publication.

Image Credits and Public Domain Notice

All images, illustrations, graphics, photographs, visual references, and related materials included in this publication are believed to be in the public domain, used under applicable licenses or permissions, created by the author or publisher, or obtained from sources that clearly identify them as available for lawful use.

The author and publisher have made a good faith effort to use only visual materials that are lawfully available for use and distribution. Where applicable, images and visual materials may have been resized, cropped, edited, enhanced, stylized, modified, reformatted, or arranged for educational, informational, design, formatting, or presentation purposes.

This publication is intended for educational, informational, and illustrative purposes only. The inclusion of any image, artwork, photograph, visual reference, symbol, historical material, design, or graphic element is not intended to claim ownership over public domain materials or any third-party intellectual property rights that remain with their respective owners.

If you believe any image or visual material included in this publication has been used in error, please contact the author or publisher so the matter may be reviewed and corrected in future editions or releases if necessary.

Any factual information, specifications, historical references, values, descriptions, statistics, or related content accompanying visual materials are provided for general informational purposes only and may change over time. Readers should conduct their own independent research and consult appropriate sources or qualified professionals when making decisions based on information presented in this publication.

This publication is independently produced and is not affiliated with, endorsed by, sponsored by, or associated with any company, organization, government agency, institution, trademark holder, or third party unless specifically stated otherwise.

Table of Contents

Introduction

Electronic waste, commonly called **e-waste**, has become one of the fastest-growing waste streams in the world. Every year, millions of computers, smartphones, tablets, printers, televisions, servers, networking devices, and countless other electronic products reach the end of their useful lives. While many people see outdated technology that belongs in storage closets, garages, or landfills, experienced recyclers recognize something entirely different. Inside those discarded devices are reusable components, repairable equipment, recyclable materials, and valuable commodities that continue to hold significant economic value long after the original owner has finished using them. Understanding how to recognize that value is the foundation of every successful e-waste recycling business.

Why E-Waste Recycling Is One of Today's Fastest-Growing Industries

Technology evolves quickly. Businesses upgrade computers every few years, schools replace aging classroom equipment, hospitals modernize medical technology, governments retire outdated electronics, and households regularly purchase newer smartphones, televisions, and laptops. As each new generation of technology enters the marketplace, another generation leaves service. This constant cycle creates an enormous supply of electronic equipment that must be managed responsibly.

Unlike ordinary household waste, electronic devices are made from dozens of different materials. A single desktop computer may contain steel, aluminum, copper, plastics, circuit boards, memory modules, processors, power supplies, storage devices, cooling systems, wiring, connectors, and other reusable parts. Smartphones, networking equipment, printers, and servers contain many of these same materials in different combinations. Learning to recognize the value of each component allows recyclers to recover far more than simply scrap metal.

One of the greatest misconceptions about e-waste is that every outdated electronic device should immediately be dismantled or shredded. In reality, many electronics continue to have value as complete products or replacement parts. A computer that no longer meets the needs of a large business may still serve a student, a small office, or a home user. A damaged laptop may contain perfectly functioning memory, processors, storage drives, or displays that can be reused in another machine. Successful e-waste businesses understand that maximizing reuse often creates greater value than immediately recycling raw materials.

The continued growth of cloud computing, artificial intelligence, data centers, consumer electronics, renewable energy systems, and connected devices means the volume of electronic waste will likely continue increasing for years to come. As more electronics enter the marketplace, the need for businesses capable of responsibly collecting, evaluating, refurbishing, and recycling those products continues to grow as well.

How This Book Will Help You Build an E-Waste Business

Starting an e-waste recycling business can appear complicated because the industry combines many different skills. There are electronics to identify, equipment to test, valuable components to recognize, data security considerations, environmental responsibilities, customer relationships to develop, and multiple ways to generate revenue from the same piece of equipment.

This guide is designed to simplify that process.

Throughout the following chapters, you'll learn how modern e-waste recycling businesses operate, where electronic waste comes from, how to build dependable collection sources, and how electronics move from collection to inspection, testing, refurbishment, dismantling, recycling, and resale. You'll discover why successful businesses inspect every device before deciding whether it should be repaired, harvested for parts, or recycled as raw material.

You'll also learn how valuable components such as processors, memory, hard drives, solid-state drives, power supplies, graphics cards, motherboards, networking equipment, and other electronic assemblies fit into the broader recycling business. Rather than viewing every electronic device as scrap, you'll begin seeing opportunities to recover value at multiple stages of the recycling process.

Whether your goal is earning supplemental income, building a part-time business, or eventually operating a larger recycling company, this guide provides the practical foundation needed to begin confidently while avoiding many of the mistakes that new recyclers often make.

Safety, Ethics, and Responsible Electronics Recycling

A successful e-waste recycling business is built on more than profitability. It is built on trust, responsibility, and professionalism. Every electronic device represents someone else's property, someone else's data, and materials that deserve to be handled safely and responsibly.

One of the most important responsibilities in the industry is **data security**. Computers, servers, laptops, smartphones, and storage devices often contain confidential personal information, financial records, business documents, customer databases, medical information, or proprietary company files. Professional recyclers treat every storage device as though it contains sensitive information until it has been properly sanitized or physically destroyed according to accepted data destruction practices.

Environmental responsibility is equally important. Many electronic devices contain rechargeable batteries, mercury-containing lamps, toner cartridges, capacitors, and other materials that require proper handling before recycling can begin. Separating these materials protects workers, prevents contamination, and helps ensure that valuable recyclable materials can be recovered safely.

Ethical sourcing is another essential principle. Electronics should always be obtained through legitimate channels such as business upgrades, public surplus sales, donations, auctions, recycling events, IT asset disposition programs, manufacturer take-back programs, and other authorized collection methods. Building a reputation for honesty, reliability, and professionalism encourages organizations to continue trusting your business with their retired equipment.

Throughout this book, these principles remain constant. Safe handling, legal compliance, customer privacy, environmental stewardship, and ethical business practices are not separate from a successful recycling business. They are the very qualities that help create one.

Last Thoughts

Starting an e-waste recycling business is about far more than collecting unwanted electronics. It is about recognizing opportunity where others see obsolete technology, extending the useful life of equipment whenever possible, recovering valuable materials responsibly, and helping create a more sustainable future. By understanding why the industry continues to grow, learning how this guide will help you develop your business, and committing yourself to safe and ethical recycling practices from the very beginning, you'll build a strong foundation for everything that follows. In the next chapter, you'll explore the e-waste recycling industry in greater detail, learn what qualifies as electronic waste, examine the many categories of devices entering the recycling stream, and discover how modern recycling businesses transform discarded electronics into valuable products, reusable components, and raw materials.

Chapter 1: Understanding the E-Waste Recycling Industry

Before collecting your first computer or dismantling your first electronic device, it's important to understand the industry you're entering. E-waste recycling is much more than scrapping old electronics for metal. Modern recycling businesses recover reusable equipment, harvest valuable components, protect sensitive data, responsibly manage hazardous materials, and recycle raw materials that can be used to manufacture new products. The more you understand how the industry operates, the easier it becomes to recognize opportunities, avoid costly mistakes, and build a business that is both profitable and environmentally responsible.

What Is E-Waste and Why It Matters

Electronic waste, or e-waste, includes virtually any electronic device that has reached the end of its useful life or is no longer wanted by its owner. This includes desktop computers, laptops, smartphones, tablets, printers, televisions, monitors, servers, networking equipment, gaming consoles, office electronics, cables, power supplies, external storage devices, and countless other products that depend on electrical or electronic components.

Unlike ordinary household waste, electronic devices contain a remarkable combination of materials. Steel provides structural support, aluminum reduces weight, copper carries electrical current, plastics form protective housings, and circuit boards connect thousands of tiny electronic components. Many devices also contain small amounts of precious metals such as gold, silver, palladium, and platinum, along with reusable electronic components that continue functioning long after the original device has been retired.

As technology advances, businesses and consumers replace electronics more frequently than ever before. A company upgrading hundreds of office computers or replacing an aging server room may generate large quantities of electronic equipment in a single project. Schools, hospitals, manufacturers, retailers, financial institutions, and government agencies also replace equipment on a regular basis. These ongoing technology upgrades create a continuous stream of electronic waste that requires responsible management.

Managing e-waste properly benefits everyone involved. Businesses can safely dispose of outdated equipment while protecting sensitive information.

Manufacturers gain access to recycled raw materials that reduce dependence on newly mined resources. Consumers benefit from refurbished electronics entering the secondary market. At the same time, responsible recycling helps reduce the amount of electronic waste entering landfills while supporting a more sustainable manufacturing economy.

Types of Electronic Waste and Their Value

Not all electronic waste is created equal. Some devices are far more valuable as complete products, while others generate greater value through component harvesting or material recovery. Learning to recognize these differences is one of the most important skills an e-waste entrepreneur can develop.

Computers and laptops are among the most versatile products entering the recycling stream. Some may require only minor upgrades or repairs before they can be resold. Others may contain valuable processors, memory modules, storage drives, power supplies, cooling systems, or motherboards that remain fully functional even if the computer itself no longer operates.

Servers and networking equipment often contain higher-value enterprise components designed for continuous operation. Businesses frequently replace this equipment because of technology upgrades rather than complete failure, creating opportunities for refurbishment, component harvesting, and resale.

Smartphones and tablets present another category of electronic waste. Even damaged devices may contain usable displays, cameras, batteries, circuit boards, charging ports, or other components that retain value. Similarly, printers, monitors, televisions, and office equipment may contain reusable parts, recyclable metals, or specialized electronic assemblies depending on their condition.

Accessories also deserve attention. Power adapters, charging cables, networking hardware, keyboards, computer mice, docking stations, speakers, and external storage devices often accumulate during technology upgrades. While individual items may appear inexpensive, large quantities can create meaningful revenue streams when properly sorted and marketed.

Understanding the strengths of each product category helps determine whether a device should be refurbished, dismantled for parts, or recycled for raw materials. Successful recyclers evaluate every item individually rather than assuming every electronic product belongs in the shredder.

The Modern E-Waste Recycling Process

Professional e-waste recycling follows a systematic process designed to recover the greatest possible value from every device while protecting people, data, and the environment.

The process usually begins with collection. Electronics arrive through business upgrades, public collection events, manufacturer take-back programs, educational institutions, surplus sales, municipalities, and individual consumers. Once received, each shipment is documented, organized, and prepared for evaluation.

The next stage involves inspection and sorting. Rather than immediately dismantling equipment, trained technicians evaluate each device to determine its condition. Some electronics may be fully functional, while others require minor repairs or cleaning before resale. Equipment that cannot be economically repaired may still contain valuable reusable components that can be harvested before recycling begins.

Data-bearing devices receive additional attention during this stage. Hard drives, solid-state drives, servers, and other storage devices must be handled carefully to protect customer privacy. Depending on the intended outcome, storage media may be securely erased using approved data sanitization methods or physically destroyed before further processing.

After reusable products and components have been removed, the remaining materials enter the recycling process. Electronics may be dismantled manually or processed using specialized industrial equipment that separates metals, plastics, circuit boards, glass, and other materials into individual recycling streams. Magnets, optical sorting systems, air separation equipment, and other technologies help recover valuable materials with remarkable efficiency.

Once separated, these materials are sold to manufacturers, refiners, smelters, plastic processors, and other industries that transform recycled materials into new products. What once appeared to be obsolete electronic waste becomes the raw material for future generations of computers, appliances, vehicles, consumer electronics, and industrial equipment.

Understanding this complete process helps explain why successful e-waste businesses focus on maximizing value at every stage. Every decision, whether refurbishing a laptop, harvesting a processor, securely erasing a hard drive, or recycling copper wiring, contributes to the overall profitability and sustainability of the business.

Last Thoughts

Understanding the e-waste recycling industry begins with recognizing that electronic waste is far more than discarded technology. Every computer, smartphone, server, printer, and electronic device represents a collection of reusable products, valuable components, recyclable materials, and future business opportunities. By learning what qualifies as e-waste, understanding the different categories of electronic equipment, and exploring how modern recycling businesses recover value from every stage of the process, you've established the foundation needed to build a successful recycling business. In the next chapter, you'll learn where electronic waste comes from, how to build reliable collection sources, and how strong supplier relationships become one of the most valuable assets in a growing e-waste recycling business.

Chapter 2: Finding and Collecting Electronic Waste

Every successful e-waste recycling business depends on one essential ingredient: a reliable supply of electronics. No matter how much you know about refurbishment, dismantling, or recycling, your business cannot grow without a consistent flow of equipment entering your operation. Fortunately, electronic waste is generated almost everywhere. Businesses upgrade computers, schools replace classroom technology, hospitals modernize equipment, retailers refresh point-of-sale systems, and homeowners regularly replace personal electronics. Learning where these products come from and building relationships with the people who generate them is one of the most valuable skills an e-waste entrepreneur can develop.

Residential and Community Collection Opportunities

Many people begin their e-waste business by collecting electronics from individuals within their local communities. Homes often contain old computers, broken printers, outdated gaming consoles, unused networking equipment, DVD players, monitors, keyboards, cables, and smartphones that have been sitting in closets, basements, garages, or storage rooms for years. While each individual collection may seem small, these opportunities can add up quickly when approached consistently.

Community recycling events are another excellent source of electronic waste. Many cities, counties, environmental organizations, and nonprofit groups organize periodic electronics collection days where residents can safely dispose of unwanted devices. Participating in these programs or partnering with event organizers can help introduce your business to potential customers while providing access to a steady supply of electronics.

Neighborhood clean-up events, moving sales, estate sales, garage sales, and community donation drives may also produce valuable opportunities. During these events, people are often focused on clearing space rather than maximizing the value of older electronics. Equipment that no longer serves one household may still contain reusable components or recyclable materials that retain significant value.

Building trust within your community is just as important as collecting equipment. Responding promptly, arriving on time, communicating clearly, and handling electronics responsibly encourages customers to recommend your services to friends, family members, and neighbors. Word-of-mouth referrals often become one of the most valuable marketing tools available to a growing recycling business.

As your reputation develops, residential collections may evolve from occasional pickups into a dependable source of inventory that supports the continued growth of your business.

Business, School, and Commercial E-Waste Sources

While residential collections provide valuable opportunities, many successful e-waste businesses eventually focus on commercial clients because businesses often replace electronics in much larger quantities. A single office technology upgrade may produce dozens or even hundreds of computers, monitors, printers, networking devices, telephones, and accessories at one time.

Businesses regularly replace electronic equipment for many reasons. Computers become outdated, software requirements change, warranties expire, security policies evolve, and organizations adopt newer technologies to improve productivity. Even when older equipment continues functioning properly, it may no longer meet the organization's operational needs.

Schools, colleges, universities, and training centers also generate significant amounts of electronic waste. Computer laboratories, classrooms, administrative offices, and libraries periodically upgrade technology, creating opportunities for responsible recyclers to collect retired equipment while supporting educational institutions with environmentally responsible disposal.

Healthcare facilities, manufacturing companies, financial institutions, law firms, retailers, municipalities, warehouses, and distribution centers all rely heavily on electronic equipment. Servers, networking hardware, barcode scanners, office computers, security systems, point-of-sale terminals, and communication equipment eventually reach retirement and require responsible disposition.

Many organizations also appreciate working with recyclers who understand the importance of **IT asset disposition (ITAD)**. Businesses are often less concerned about receiving the highest possible scrap price than they are about ensuring equipment is handled securely, data is protected, and electronics are managed in an environmentally responsible manner. Providing dependable service and maintaining professionalism can help establish long-term relationships that generate recurring business.

Commercial accounts often become the backbone of successful e-waste recycling businesses because they provide predictable inventory, larger collection volumes, and ongoing opportunities as organizations continue upgrading their technology.

Building Collection Partnerships and Supply Networks

Long-term success in the e-waste industry is rarely built through one-time collections. Instead, it comes from creating a reliable network of people and organizations that regularly generate electronic waste and trust your business to handle it responsibly.

Developing partnerships begins with understanding the needs of potential suppliers. Businesses want dependable collection services that minimize disruption. Schools want responsible disposal methods that protect student information. Government agencies require compliance with established procedures. Nonprofit organizations often appreciate recyclers who can help remove outdated equipment while supporting environmental sustainability.

Building these relationships takes time. Introducing your business professionally, explaining your services clearly, following through on commitments, and maintaining regular communication help establish credibility. Over time, organizations begin viewing your business as a dependable recycling partner rather than simply someone collecting unwanted electronics.

Networking within related industries also creates valuable opportunities. Computer repair shops, managed IT service providers, office equipment dealers, moving companies, property managers, liquidation firms, office furniture installers, and commercial clean-out companies frequently encounter electronic equipment that requires responsible disposition. Establishing referral relationships with these businesses can create a steady flow of new inventory while benefiting everyone involved.

As your supply network expands, organization becomes increasingly important. Maintaining records of collection dates, supplier contacts, equipment categories, pickup schedules, and customer preferences helps improve efficiency while strengthening long-term relationships. Businesses appreciate consistency, and reliable service often leads to repeat collections year after year.

The strongest e-waste businesses are built on trust. Organizations entrust recyclers with valuable equipment, sensitive information, and environmental responsibilities. By honoring that trust through professionalism, transparency, and ethical business practices, you create a reputation that continues generating opportunities long into the future.

Last Thoughts

Finding electronic waste is not about waiting for opportunities to appear. It is about understanding where electronics naturally reach the end of their service lives and building relationships with the people and organizations responsible for replacing them. Residential collections, community programs, commercial clients, schools, government agencies, and business partnerships all contribute to a dependable supply of equipment that supports a growing recycling business. In the next chapter, you'll learn how to inspect incoming electronics, evaluate their condition, test equipment for reuse, harvest valuable components, and protect sensitive data before determining the best path for each device.

Chapter 3: Sorting, Testing, and Harvesting Electronics

Once electronic equipment has been collected, the real work begins. Every computer, printer, server, smartphone, monitor, or networking device represents a decision. Should it be cleaned and resold? Can it be repaired? Does it contain valuable reusable components? Or has it reached the point where recycling is the most practical option? Successful e-waste businesses answer these questions through a systematic evaluation process rather than making assumptions based solely on appearance. Careful inspection, organized sorting, responsible data protection, and thoughtful component harvesting help maximize the value of every device while reducing unnecessary waste.

Inspecting and Sorting Incoming Electronics

The first step after receiving electronic equipment is conducting a careful inspection. Rather than immediately dismantling every device, experienced recyclers evaluate each item individually to determine its condition and its potential value. This process often begins with recording basic information such as the manufacturer, model number, condition, and whether the equipment appears complete.

Visual inspection reveals a surprising amount about an electronic device. Cracked screens, missing components, water damage, broken ports, severe corrosion, or obvious physical abuse may indicate that repair is unlikely to be economical. At the same time, many electronics that appear outdated or dusty remain fully functional after simple cleaning and testing.

Organization is equally important during this stage. Keeping laptops together, desktop computers together, printers together, networking equipment together, and accessories in their own designated areas makes inventory management far more efficient. It also helps technicians process similar products using consistent workflows instead of constantly switching between different types of equipment.

Many businesses also separate equipment into broad categories such as products intended for resale, devices requiring additional testing, equipment suitable for parts harvesting, and electronics destined for recycling. This organized approach reduces confusion while allowing valuable products to move efficiently through the business.

As inventory grows, proper labeling becomes increasingly valuable. Asset tags, inventory numbers, collection dates, customer information, and processing status all help track equipment throughout the recycling process. Good organization not only improves efficiency but also creates a more professional operation capable of handling larger commercial projects.

Testing, Refurbishing, and Harvesting Valuable Components

After inspection comes evaluation. The goal is to determine whether a device can continue serving its original purpose or whether its greatest value lies within its individual components.

Many computers and laptops require only minor work before becoming usable again. Cleaning dust from cooling systems, replacing storage drives, upgrading memory, installing a fresh operating system, or replacing inexpensive components can often restore equipment to dependable working condition. Refurbished electronics frequently provide greater value than dismantling them for scrap because complete, functioning systems appeal to businesses, students, schools, and home users.

When complete refurbishment is not practical, harvesting components often becomes the next best option. A damaged laptop may still contain perfectly functional memory modules, processors, cooling fans, wireless network cards, keyboards, displays, speakers, webcams, or solid-state drives. Likewise, desktop computers may provide reusable power supplies, graphics cards, processors, expansion cards, optical drives, and motherboards.

Servers and enterprise equipment deserve especially careful evaluation because many of their components retain significant value in secondary markets. Enterprise processors, high-capacity memory modules, network interface cards, power supplies, and storage systems are often reused long after the original server has been retired.

Testing components before resale helps protect both your customers and your reputation. Memory can be verified using diagnostic software, storage drives can be checked for health and remaining lifespan, power supplies can be tested with dedicated testers, and complete systems can undergo hardware diagnostics before being offered for sale. Selling properly tested equipment reduces returns while building confidence in your business.

Learning to recognize when equipment should remain intact, when it should be refurbished, and when components should be harvested is one of the most valuable skills an e-waste entrepreneur can develop. Every thoughtful decision increases the amount of value recovered from incoming electronics.

Data Security and Hazardous Material Removal

Protecting customer information is one of the greatest responsibilities in the e-waste recycling industry. Every computer, laptop, server, smartphone, tablet, external drive, or storage device should be treated as though it contains confidential information until proven otherwise.

Hard drives and solid-state drives frequently store financial records, personal photographs, business contracts, customer databases, medical records, passwords, emails, and countless other forms of sensitive information. Professional recyclers establish clear procedures for securely sanitizing storage devices before resale or physically destroying storage media when reuse is not appropriate.

Documenting these processes also becomes important, particularly when working with commercial clients. Many businesses require written confirmation that storage devices have been properly erased or destroyed before equipment leaves the recycling facility. Developing consistent procedures helps demonstrate professionalism while strengthening customer confidence.

Beyond data security, electronics often contain materials requiring careful handling before dismantling or recycling begins. Rechargeable batteries, toner cartridges, mercury-containing lamps, capacitors, and other specialized components should be removed before equipment enters mechanical recycling processes. Proper separation protects workers, reduces environmental risks, and improves the efficiency of downstream recycling operations.

Clean workspaces also contribute to safer operations. Keeping hazardous materials separated from reusable inventory, organizing removed components, and maintaining designated processing areas reduces confusion while creating a more efficient workflow. As your business grows, these organizational habits become increasingly valuable.

Ultimately, protecting sensitive information and responsibly managing hazardous materials are not simply regulatory obligations. They are essential parts of building a business that customers trust. Organizations are far more likely to become repeat clients when they know their retired equipment will be handled professionally from the moment it arrives until every reusable component has been recovered and every recyclable material has been processed responsibly.

Last Thoughts

Sorting, testing, and harvesting electronics form the foundation of a successful e-waste recycling business. Careful inspection, organized inventory management, thoughtful refurbishment, responsible component harvesting, secure data handling, and proper hazardous material removal all contribute to maximizing the value of every device while protecting customers and the environment. Rather than viewing electronics as disposable products, successful recyclers evaluate each item for its highest and best use before deciding how it should move through the recycling process. In the next chapter, you'll learn how electronics are dismantled and processed for material recovery, explore the different ways recyclers generate revenue from refurbished equipment, reusable parts, and scrap materials, and gain a better understanding of the valuable commodities hidden inside everyday electronic devices.

Chapter 4: Processing and Selling Electronic Materials

Once reusable devices have been identified, valuable components have been harvested, and sensitive data has been protected, the remaining electronics move into the final stages of the recycling process. At this point, the objective shifts from preserving complete products to recovering as much value as possible from the materials that remain. Successful e-waste businesses understand that profitability does not come from a single revenue source. Instead, it comes from combining refurbished electronics, reusable components, recyclable materials, and strong relationships with buyers who specialize in different parts of the recycling industry. Learning how these materials are processed and sold allows you to make informed decisions that maximize the value of every load of electronic waste.

Dismantling Electronics and Material Recovery

Not every electronic device can be repaired or refurbished. Some products are too heavily damaged, too outdated, or too expensive to restore.

When that happens, the next step is recovering the individual materials and components that still have value.

Professional recyclers typically begin by removing components that require special handling. Rechargeable batteries, toner cartridges, mercury-containing lamps, capacitors, and other potentially hazardous materials are separated before additional processing begins. This protects workers, prevents contamination, and helps ensure that the remaining materials can be processed safely.

Once hazardous materials have been removed, electronics may be dismantled by hand or processed using specialized equipment, depending on the size and nature of the operation. Manual dismantling allows recyclers to recover reusable components, wiring, power supplies, cooling systems, processors, memory modules, and circuit boards before the remaining materials enter mechanical processing.

Larger recycling facilities often use industrial shredders to reduce electronics into smaller pieces. After shredding, various separation technologies recover different material streams. Powerful magnets remove steel and iron, while other equipment separates aluminum, copper, plastics, glass, and printed circuit boards. Advanced recycling facilities may also use optical sorting systems, density separation, and infrared technology to improve the purity of recovered materials.

The goal is not simply to destroy electronics. It is to separate valuable materials efficiently so they can reenter the manufacturing process as raw materials for new products.

Selling Refurbished Electronics, Parts, and Scrap Materials

One of the greatest advantages of the e-waste industry is that a single shipment of electronics can produce multiple sources of revenue. Rather than relying only on scrap metal prices, successful businesses evaluate every item to determine its highest potential value before deciding how it should be sold.

Complete refurbished computers, laptops, servers, networking equipment, monitors, and business electronics often generate the highest returns. Organizations replacing large quantities of equipment frequently retire devices because of upgrade schedules rather than complete failure. After cleaning, testing, repairing, and upgrading these products, they may be suitable for resale to individuals, schools, nonprofits, small businesses, or international markets.

Equipment that cannot be sold as complete systems may still provide valuable replacement parts. Processors, memory modules, solid-state drives, hard drives, graphics cards, power supplies, cooling fans, motherboards, network interface cards, and other components are regularly purchased by repair businesses, computer technicians, and electronics enthusiasts seeking replacement parts.

After reusable equipment and components have been removed, the remaining materials continue generating value through recycling. Circuit boards, copper wiring, aluminum heat sinks, steel cases, brass connectors, electric motors, cables, plastics, and other materials are sorted into separate categories before being sold to recyclers, refiners, processors, or manufacturers.

Diversifying your revenue streams helps reduce dependence on any single market. If refurbished computer prices decline, component sales or scrap materials may continue generating income. Likewise, changes in metal prices may be balanced by increased demand for refurbished electronics or replacement parts. Building multiple income streams creates a more resilient business capable of adapting to changing market conditions.

Understanding Precious Metals and Other Valuable Commodities

Many people are first attracted to e-waste because they hear that electronics contain gold. While this is true, gold represents only one small part of the overall value contained within modern electronic devices.

Circuit boards often contain small amounts of gold, silver, palladium, and other precious metals used to improve electrical conductivity and corrosion resistance. Connectors, processors, memory modules, and certain expansion cards may also contain precious metal plating depending on their design and intended application. Although the amount contained in any single device is usually small, processing large quantities of electronics allows professional refiners to recover these valuable materials efficiently.

Copper is another important commodity found throughout electronic equipment. Wiring, transformers, electric motors, power supplies, coils, connectors, and circuit boards all contain copper in varying amounts. Aluminum is commonly found in heat sinks, structural components, and electronic housings, while steel provides strength for computer cases, server racks, and many internal assemblies.

Plastics also play an important role in modern electronics recycling. Computer housings, monitor bezels, keyboards, printers, and countless other products contain recyclable plastics that can be processed and manufactured into new products after proper sorting and cleaning.

Understanding these different material streams helps explain why successful e-waste businesses rarely focus on only one commodity. A shipment of electronics may generate income from refurbished computers, harvested components, circuit boards, copper, aluminum, steel, plastics, cables, and precious metals, all from the same collection. Looking at electronic waste as a combination of valuable resources rather than a single product allows recyclers to recover far greater value over time.

Last Thoughts

Processing and selling electronic materials is about recovering value at every possible stage. From refurbished computers and reusable components to recyclable metals, plastics, and precious metals, every electronic device offers multiple opportunities when evaluated thoughtfully. The most successful recyclers understand that maximizing value requires patience, organization, and a willingness to consider every possible use before sending materials to the recycling stream. In the next chapter, you'll learn how to establish the operational side of your business, choose appropriate tools and equipment, develop efficient workflows, and market your services to build long-term relationships with customers and suppliers.

Chapter 5: Starting and Growing Your E-Waste Business

Starting an e-waste recycling business does not require building a massive industrial recycling facility on your first day. Many successful businesses begin with a small workspace, basic tools, and a commitment to learning the industry one step at a time. As experience grows, so do the services offered, the number of collection partners, and the volume of electronics being processed. Building a successful business is less about starting large and more about creating efficient systems that can grow over time. Careful planning, thoughtful investments, and strong customer relationships often have a greater impact on long-term success than expensive equipment purchased too early.

Planning Your Business and Workspace

Every successful business begins with a plan. Before collecting large quantities of electronics, it is important to decide what type of e-waste business you want to build. Some businesses specialize in collecting and reselling refurbished computers. Others focus on harvesting valuable components, while larger operations concentrate primarily on processing electronic scrap for recycling. Many successful companies eventually combine all three approaches.

Choosing an appropriate workspace is one of the first major decisions. A small operation may begin in a garage, workshop, warehouse, or commercial storage unit, provided local zoning regulations and business requirements allow it. As inventory grows, additional space becomes necessary for receiving incoming electronics, inspecting equipment, testing devices, storing reusable inventory, organizing recyclable materials, and preparing outgoing shipments.

Organization should be incorporated into the workspace from the very beginning. Creating designated areas for incoming equipment, products awaiting testing, refurbished inventory, harvested components, recyclable materials, and hazardous items improves efficiency while reducing confusion. Employees and customers alike benefit from a clean, organized environment where equipment can be located quickly and processed consistently.

Planning for future growth is equally important. Even if your business begins with only a few computers each week, establishing organized systems early makes expansion much easier. Simple habits such as inventory tracking, equipment labeling, customer records, and organized storage become increasingly valuable as the business continues to grow.

Equipment, Tools, and Operational Considerations

The equipment needed for an e-waste business depends largely on the services you plan to offer. A business focused primarily on collecting and reselling computers requires different tools than one specializing in industrial recycling.

For most beginners, a basic toolkit is sufficient. Standard hand tools such as screwdrivers, precision screwdriver sets, pliers, wire cutters, anti-static wrist straps, flashlights, cable organizers, and storage containers allow many electronics to be inspected and disassembled safely. Computer diagnostic software, storage drive testing tools, memory testing utilities, and operating system installation media also become valuable when refurbishing computers and laptops.

As the business expands, additional equipment may include shelving systems, workbenches, barcode scanners, inventory software, pallet jacks, secure storage cabinets, testing stations, and packaging materials for shipping refurbished products or electronic components. Businesses that process larger quantities of electronic scrap may eventually invest in balers, cable stripping equipment, forklifts, or other specialized material handling equipment.

Operational procedures are just as important as physical tools. Developing consistent workflows helps every electronic device move efficiently through the business. Equipment should be logged upon arrival, inspected, tested, categorized, and assigned to the appropriate processing path. Whether a device is refurbished, harvested for parts, or recycled, following the same process each time improves productivity while reducing mistakes.

Safety should remain a priority throughout every stage of operation. Employees should understand proper lifting techniques, safe tool usage, electrical safety practices, battery handling procedures, and data security responsibilities. A safe workplace not only protects people but also creates a more professional business environment.

Marketing Your Services and Building Long-Term Customers

Even the most organized recycling business cannot grow without customers. Fortunately, electronic waste is generated continuously by businesses, schools, healthcare organizations, manufacturers, retailers, government agencies, and households. The challenge is making sure those organizations know your business exists and understand the value you provide.

A professional online presence is an excellent place to begin. A well-designed website should clearly explain your services, collection procedures, accepted electronics, contact information, service area, and commitment to responsible recycling. Including information about refurbishment, secure data destruction, business pickups, and environmental responsibility helps potential customers understand what distinguishes your business from competitors.

Local marketing also plays an important role. Networking with computer repair shops, managed IT service providers, office equipment dealers, commercial property managers, schools, nonprofit organizations, municipalities, and business associations helps establish valuable referral relationships. Many organizations regularly encounter electronic equipment that requires responsible disposal but may not have an established recycling partner.

Providing exceptional customer service is one of the most effective marketing strategies available. Returning phone calls promptly, arriving on schedule, communicating clearly, respecting customer property, and following through on commitments encourage customers to recommend your business to others. Positive experiences often lead to repeat collections, referrals, and long-term business relationships.

Educational marketing can also help establish your expertise. Sharing information about electronics recycling, environmental responsibility, secure data handling, and refurbishment through blog articles, videos, workshops, or community events demonstrates knowledge while helping potential customers better understand the value of professional e-waste recycling.

As your reputation grows, repeat customers often become your greatest source of new business. Organizations that trust your professionalism are far more likely to contact you whenever additional electronics become available, creating a steady stream of inventory that supports long-term growth.

Last Thoughts

Building an e-waste recycling business requires much more than collecting discarded electronics. Success comes from careful planning, organized operations, appropriate tools, consistent procedures, and strong customer relationships. By creating an efficient workspace, investing in the right equipment as your business grows, and developing a reputation for professionalism and reliability, you establish a business capable of serving customers for many years. In the next chapter, you'll explore how to scale your e-waste recycling business, expand into new revenue streams, adapt to changing technologies, and build a sustainable company positioned for long-term success.

Chapter 6: Scaling Your E-Waste Recycling Business

Building an e-waste recycling business is only the beginning. As your experience grows, you'll begin identifying new opportunities to expand your services, increase efficiency, and generate additional revenue from the same stream of electronic equipment. Growth does not always mean processing more electronics. In many cases, it means recovering more value from every device that enters your business. Companies that continue learning, adapting to new technologies, and strengthening relationships with customers are often the ones that achieve long-term success in this rapidly evolving industry.

Expanding Services and Increasing Revenue Streams

One of the greatest strengths of an e-waste recycling business is the variety of services it can provide. Many companies begin by collecting unwanted electronics but eventually expand into refurbishment, data destruction, IT asset disposition, component harvesting, equipment resale, asset recovery, and consulting services.

As your knowledge increases, you may choose to specialize in particular types of electronics. Some businesses focus on enterprise servers and networking equipment, while others become experts in computer refurbishment, smartphone recovery, business technology upgrades, or industrial electronics. Developing expertise in a specific area often helps distinguish your business from competitors while allowing you to command higher prices for specialized services.

Offering secure pickup services is another way to increase value for customers. Many businesses prefer working with recyclers who can collect equipment directly from their facilities, document the transfer of assets, and provide clear communication throughout the recycling process. Making the recycling experience convenient encourages repeat business while strengthening long-term customer relationships.

Additional revenue opportunities may also come from offering data destruction services, equipment audits, inventory management, decommissioning projects, office technology clean-outs, warehouse liquidations, and surplus asset recovery. These services often complement electronics recycling while helping businesses responsibly manage retired equipment.

As your operation expands, remember that every new service should support your overall business goals. Sustainable growth comes from adding services thoughtfully rather than attempting to offer everything at once.

Industry Trends and Future Opportunities

Technology never stands still, and neither does the e-waste recycling industry. Every advancement in consumer electronics, cloud computing, artificial intelligence, renewable energy, automation, and telecommunications creates new opportunities for businesses capable of adapting to change.

Businesses continue upgrading computers, servers, networking equipment, storage systems, and communication technologies to improve performance and security. At the same time, consumers regularly replace smartphones, tablets, wearable devices, gaming systems, smart home products, and entertainment electronics. As these technologies evolve, the volume of electronic equipment entering the recycling stream continues to grow.

The increasing importance of sustainability has also influenced how organizations manage electronic assets. Many businesses now include responsible recycling as part of their environmental goals and seek recycling partners that can document proper handling, maximize equipment reuse, and minimize waste. Companies that demonstrate professionalism, transparency, and environmental responsibility are often well positioned to benefit from this growing demand.

The continued growth of data centers, cloud infrastructure, and enterprise computing also creates opportunities for recyclers specializing in commercial technology. Large organizations regularly replace servers, storage equipment, networking hardware, and supporting infrastructure as part of planned technology refresh cycles. Understanding these industries can open doors to larger projects and long-term commercial relationships.

Remaining curious about emerging technologies allows your business to evolve alongside the industry. The electronics being recycled ten years from now will not be identical to today's equipment, but the principles of responsible collection, careful evaluation, secure data handling, refurbishment, and material recovery will remain just as important.

Building a Sustainable Recycling Business

Long-term success depends on building systems that allow your business to operate consistently regardless of changes in technology or market conditions. Sustainable businesses focus on efficiency, professionalism, customer relationships, and continuous improvement rather than relying on short-term opportunities.

Organization remains one of your greatest competitive advantages. Accurate inventory records, documented processing procedures, organized storage systems, and dependable scheduling create smoother operations while making it easier to manage larger volumes of equipment. As your business grows, these systems become increasingly valuable.

Investing in your knowledge is equally important. Learning about new electronic devices, recycling technologies, refurbishment techniques, data security standards, and environmental regulations helps ensure your business remains competitive as the industry continues evolving. Businesses that continue learning are often the first to recognize new opportunities.

Strong partnerships also contribute to long-term stability. Relationships with businesses, schools, government agencies, manufacturers, IT service providers, computer repair shops, wholesalers, refurbishers, recyclers, and downstream processors create a dependable network that supports growth over many years. Trust develops through consistent service, honest communication, and professional conduct.

Perhaps the most valuable investment you can make is your reputation. Customers trust you with expensive equipment, confidential information, and environmentally sensitive materials. Protecting that trust through ethical business practices, secure data handling, responsible recycling, and excellent customer service creates a business that people are confident recommending to others.

Successful e-waste businesses are not built overnight. They grow steadily through careful planning, strong relationships, continuous learning, and a commitment to doing things the right way. Those qualities continue creating opportunities regardless of how technology changes in the years ahead.

Last Thoughts

Scaling an e-waste recycling business is about much more than processing larger quantities of electronics. It is about expanding your services, identifying new revenue opportunities, adapting to changing technologies, and building systems capable of supporting long-term growth. By remaining committed to professionalism, environmental responsibility, customer service, and continuous improvement, you position your business to succeed in one of the world's fastest-growing industries. As electronic technology continues evolving, businesses that combine technical knowledge with ethical practices and dependable service will remain well prepared to meet the challenges and opportunities of the future.

About the Author

Roggie "Raj" Clark is an entrepreneur and a 10-year marketing professional with a proven track record in e-commerce and digital business growth. Having built and managed successful online stores with thousands of sales and consulted for numerous marketing clients, Roggie has mastered the intricacies of online marketing, branding, and consumer behavior. He has also written several popular books on digital marketing, SEO (search engine optimization), and programming (HTML, CSS, and JavaScript).

As a former marketing manager at a fintech company—a role he secured through his own marketing and financial expertise—Roggie gained firsthand experience in financial strategy and wealth-building. This journey inspired him to share insights with amateur investors, helping them save money, make smarter financial decisions, and build long-term wealth. Raj currently lives in Maryland with his sister and niece.